Ages of stars

STARS—THEIR AGES, LIFESPANS AND TIMES TILL THEIR DEATHS

JOHN HEBERT
3/22/18-5/25/18;
12/26-28/18;1/3/19

for-meryem,cathy,
bunny

The universe is very big. If our earth were the size of a smoke particle, the nearest star would be located 2 miles away. The distance spanning across the universe would be equivalent to going the distance of 3.875 million times around the earth. Astronomically, time is very long. If an inch were to equal a year, than a rather short lived star that lives 100 million years would be the distance of 1,578 miles.

Our sun's lifespan would be a distance of going 6.31 times around the earth (75 years=1 inch:

100 million years=21 miles), and there are stars whose lifetime distances would be a distance of going 6,310 times around the earth. The number of

stars in the universe is

very large. If a star were equal in size to a grain of sand, then all the stars in the universe would make a compact ball 12 miles large. Each person on earth receiving an equal number of stars from the total number of them would each have a ball of stars 40 feet in diameter. Stars can be very large. If the earth were an inch large, the largest of star would *have diameters 2.8 miles large. About infinite timespan— here is my incomplete attempt to describe infinity using our immortalness as souls. Let us say the each atom were equal to a human lifespan. There are 525 billion atoms which make up a smoke particle. There are also 240 million of these which spans 1 inche's length.*

There are 1.6 billion of these inches needed to go around the earth. There are 3.1 billion times *that distance needed to go to the nearest star. Then, there are 21.6 billion of those distances needed to span the distance across the universe. If we were to add up all of the total of our lifetimes (the atom's being a lifetime),our souls and consciousnesses will be experiencing life of some kind life during all that time, and that total time is just the beginning and really like being equal to zero time because eternity never ends. We will always be alive and conscious because we are immortal. I wonder what we will be doing during time. I feel it will not be static existence but rather one of growth centering on love.*

BROWN DWARF STAR

Stars name— upsilon andromedae d (majriti)

*Right ascension—
01 hours 36
minutes 48
seconds*

*declination— 41
degrees 24
minutes 20
seconds*

Stellar class—orbiting F8(5) star

constellation—Andromeda

Apparent magnitude—approximately 4.68

Absolute magnitude—

approximately 4.03 distance— 43.9 light years

mass— 10.25 jupiters

radius— 1.02 jupiters

luminosity— approximately 2.08 suns

Semimajor axis— 2.53 AU

eccentricity— .299

inclination— 23.8 degrees

Period of orbit—1,276.46 days

Orbital distance—2.3 AU (214.17 million miles)

Surface temperature— less than 1,000 degrees kelvin

Velocity of motion — no information found

Movement in arcseconds/year — no information found

Direction on the unit circle — no information found

parallax — 74.26 milliarcseconds

Size in arcseconds — .003734 arc seconds

Gravitational acceleration — 80.52 feet/ second^2

Escape velocity— 36.9 miles/ second

age— less than 3 billion 120 million years

lifespan— 3.55 x 10^17 years

Main sequence (prime of its life) lifespan— never made it to the main sequence

Time left as a main sequence star — n/a

When will this star die? — 3.55 x 10^17 years

Fate of the star — black dwarf (?)

VERY YOUNG STARS

Star's name— stars in the orion nebula trapezium

Our destination tonight will be the bright inner core of the Orion Nebula (circled) centered on the Trapezium multiple star.

Stellar class— O main sequence stars

color— blue

constellation— orion

Apparent magnitude— 2.57/-.6141

Absolute magnitude— -5.51/-8.09

distance— 1,344 light years

mass— 15/30 suns

radius— 4.42/9.98 suns

luminosity— 13,701/147,885 suns

Surface temperature— 30,251/36,500 degrees kelvin

Velocity of motion — no information found

Movement in arcseconds/year — no information found

Direction on the unit circle— no information found

parallax—2.426 milliarcseconds

Size in arcseconds —.0107/.02421 arc seconds

Gravitational acceleration — 1,243.44/267.7 feet/ second^2

Escape velocity— 499.38/470 miles/ second

age— 300,000 years old

lifespan—11 million 476 thousand/ 2 million 29 thousand years

Main sequence (prime of its life) lifespan— 10 million 328 thousand/1 million 826 thousand years

Time left as a main sequence star — 10 million 28 thousand/729 thousand years

When will this star die?— 11 million 176 thousand/ 1 million 729 thousand years

Fate of the star— supernovas, neutron star/ black hole

MAIN SEQUENCE STAR'S

Stars name— tu muscae

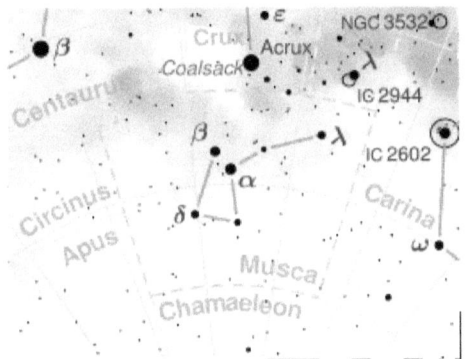

Stellar class— O7(5)+O8(5)

Color — blue constellation— musca

Right ascension— 11 hours 31 minutes 10.92470 seconds

declination— -65 degrees 44 minutes 32.1019 seconds

Apparent magnitude— 8.37

Absolute magnitude— -4.81

distance — 163 light years

mass — 16.8/10.5 suns

radius — 7.2/7.7 suns

luminosity — 105,000/35,000 suns

Surface temperature— 35,000/31,366 degrees kelvin

Velocity of motion — 2.48 miles/second approach

Movement in arcseconds/year—

.00121 seconds/year

Direction on the unit circle— 258.49 degrees

parallax—.02 milliarcseconds

Size— .3726/.398 arcseconds

Gravity— 292.78/157.34 feet/ second^2

Escape velocity— 414.08/316.55 miles/ second

age— less than 7 million 780 thousand/less than

25 million 123 thousand years lifespan— 8 million 644 thousand/ 27 million 992 thousand years Main sequence (prime of its life) lifespan— 7 million

780 thousand/ 25 million 123 thousand years

Time left on main sequence— less than 7 million 780 thousand/less than 25 million 780 thousand years

When will this star die?— greater than 864 thousand/ greater than 2 million 799 thousand years

Fate of the stars— supernovas, neutron stars

Star's name— zeta ophiuci

constellation— Ophiucus

Right ascension— 16 hours 37 minutes 09.53905 seconds

declination— -10 degrees 34

minutes 01.5295 seconds

Apparent magnitude— 2.569

Absolute magnitude— -4.2

distance—366 light years

mass— 20 suns

radius— 8.5 suns

luminosity— 91,000 suns

Surface temperature— 34,000 degrees kelvin

Rotation velocity— 248 miles/ second

Velocity of motion — 9.3 miles/second approach

Movement in arcseconds/year — .0291 seconds/year

Direction on the unit circle— 64.87 degrees

parallax—8.91 milliarcseconds

Size— .0757 arcseconds

Gravity— 292.78/157.34 feet/ second^2

Escape velocity— 414.08/316.55 miles/ second

age— 3 million old

lifespan— 55

million 902 thousand years

Main sequence (prime of its life) lifespan— 50 million 312 thousand years

Time left as a main sequence star —

47 million 312 thousand years

When will this star die?— 52 million 902 thousand years

Fate of the star— supernova, neutron star

Star's name— sigma orionis

σ Orionis (lower right) and the Horsehead nebula. The brighter stars are Alnitak and Alnilam.

Stellar class— O9.5(5) Color–

blue constellation — orion

Right ascension— 05 hours 38 minutes 42 seconds

declination— -02 degrees 36

minutes 00 seconds

Apparent magnitude— 4.07

Absolute magnitude— -3.49

distance—387.51 light years

mass— 18 suns

radius— 5.6 suns

luminosity— 41,700 suns

Surface temperature— 35,000 degrees kelvin

Velocity of motion — 18.259 miles/second approach

Movement in arcseconds/year — no information found

Direction on the unit circle— no information found

parallax—3.04 milliarcseconds

Size— .04711 arcseconds

Gravity—510.5 feet/ second^2

*Escape velocity—
486.01 miles/
second*

*age— less than 6
million 547
thousand old*

*lifespan— 3 million
years*

Main sequence (prime of its life) lifespan— 6 million 547 thousand years

Time left as a main sequence star — 3 million 547 thousand years

When will this star die?— 4 million 275 thousand years

Fate of the star— supernova, neutron star

Star's name— mu columbae

Stellar class—O7.5(5) Color—blue

constellation—colombae

Right ascension—05 hours 45 minutes 59.89496 seconds

declination— -32 degrees 18 minutes 23.1630 seconds

Apparent magnitude— 5.18

Absolute magnitude— -3.64

distance — 1,300 light years

mass — 16 suns

radius — 6.58 suns

luminosity — 45,700 suns

Surface temperature —

33,000 degrees kelvin

Velocity of motion — 67.58 miles/second recession

Movement in arcseconds/year — .02224 acrseconds/year

Direction on the unit circle— 359.98 degrees

parallax—2.51 milliarcseconds

Size— .0165 arcseconds

Gravity—281.42 feet/ second^2

*Escape velocity—
422.71 miles/
second*

*age— 2-4 million
years old*

*lifespan— 9 million
766 thousand
years*

Main sequence (prime of its life) lifespan— 8 million 789 thousand years

Time left as a main sequence star — 4 million 789 thousand-6 million

789 thousand years

When will this star die?— 5 million 766 thousand-7 million 766 thousand years

Fate of the star—supernova, neutron star

Was located in the orion nebula's trapezium 2 million 500 thousand years ago.

Star's name— acrab beta scorpi

Stellar class—B1(5)

color— blue

constellation—scorpius

Apparent magnitude— 2.62

Absolute magnitude— -3.92

distance— 400 light years

mass— 15 suns

radius— 6.3 suns

luminosity— 31,600 suns

Surface temperature— 28,000 degrees kelvin

Velocity of motion — .63 miles/ second approach

Movement in arcseconds/year—

.024596 seconds/year

Direction on the unit circle— 186.56 degrees

parallax—8.07 milliarcseconds

Size in arcseconds — .0508

Gravitational acceleration — 335.8 feet/second^2

Escape velocity — 435.78 miles/second

age— 9-12 million years old (closer to 6 million years old)

lifespan— 11 million 476 thousand years

Main sequence (prime of its life) lifespan— 10

million 328 thousand years

Time left as a main sequence star — 4 million 328 thousand years

When will this star die?— 7 million

476 thousand years
Fate of the star—supernova, neutron star

Star's name—
alpha crucis

Stellar class—
B1(5)

color— blue

constellation— crux

Apparent magnitude— 1.75

Absolute magnitude— -2.7

distance— 320 light years

mass— 15.52 suns

radius— 1.41 suns

luminosity— 1,028.37 suns

Surface temperature— 28,000 degrees kelvin

Velocity of motion — 6.944 miles/second approach

Movement in arcseconds/year — .03879 seconds/year

Direction on the unit circle— 244.97 degrees

parallax—10.1 milliarcseconds

Size in arcseconds — .01424

Gravitational acceleration — 1.314 miles/second^2

Escape velocity — 899.36 miles/second

age— 10 million 800 thousand years old

lifespan— 10 million 538 thousand years

Main sequence (prime of its life) lifespan— 9

million 485 thousand years

Time left as a main sequence star — should still be on main sequence, but age of star is more than the main sequence lifetime

When will this star die?— greater than 1 million 54 thousand years

Fate of the star— supernova, neutron star

*Star's name—
nunki
sigma sagittarius*

Sigma Sagittarii

Stellar class— B2.5(5)

color— blue

constellation— sagittarius

Apparent magnitude— 2.05

Absolute magnitude— -2.17

distance— 228 light years

mass— 7.8 suns

radius— 4.5 suns

luminosity— 3,300 suns

Surface temperature— 18,890 degrees kelvin

Velocity of motion — 6.944 miles/ second approach

Movement in arcseconds/year—

.05344 seconds/year

Direction on the unit circle— 277.58 degrees

parallax— 14.32 milliarcseconds

Size in arcseconds — .0644

Gravitational acceleration —342 feet/ second^2

Escape velocity— 356.89 miles/ second

age— 31-31.8 million years old

lifespan—58 million 852 thousand years

Main sequence (prime of its life) lifespan— greater than 52 million 967 thousand years

Time left as a main sequence star — less than 21 million 167 thousand-21 million 967 million years

When will this star die?— more than 27 million 52-27

*million 852
thousand years*

*Fate of the star—
planetary nebula,
white dwarf*

Star's name—furud zeta canis majoris

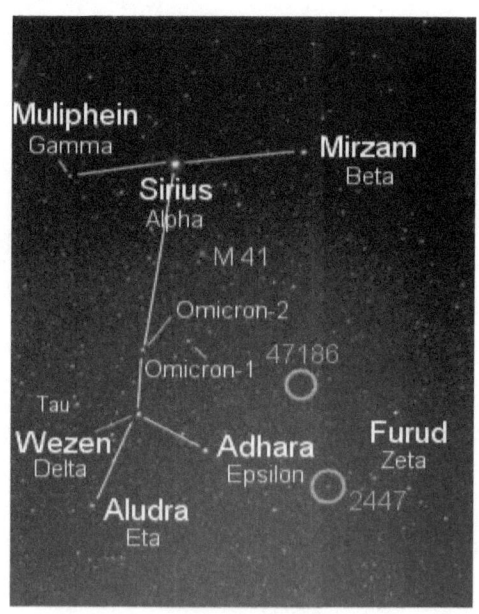

Stellar class— B2.5(5)

color— blue

constellation— canis major

Apparent magnitude— 3.025

Absolute magnitude— -2.21

distance— 362 light years

mass— 7.7 suns
radius— 3.9 suns
luminosity— 3,603 suns

Surface temperature—

18,700 degrees kelvin

Velocity of motion — 1.984 miles/second recession

Movement in arcseconds/year— .00403 seconds/year

Direction on the unit circle — 32.04 degrees

parallax — 9 milliarcseconds

Size in arcseconds — .0351

Gravitational acceleration —

449.86 feet/second^2

Escape velocity— 390.9 miles/second

age— 31.6-32.4 million years old

lifespan—60 million 782 thousand years

Main sequence (prime of its life) lifespan— 54 million 704 thousand years

Time left as a main sequence star — 22 million 300 thousand-23 million 100 thousand years

When will this star die?— million 25 million 150

thousand-28 million 400 thousand years

Fate of the star— planetary nebula, white dwarf

*Star's name—
peacock alpha
pavonis*

*Stellar class—
B3(5)*

color— blue

constellation— pavonis

Apparent magnitude— 1.94

Absolute magnitude—-1.762

distance— 179 light years

mass— 5.91 suns

radius— 4.33 suns

luminosity— 2,200 suns

Surface temperature— 17,700 degrees kelvin

Velocity of motion — 1.24 miles/second recession

Movement in arcseconds/year—

.00719 seconds/year

Direction on the unit circle — 265.1 degrees

parallax — 18.24 milliarcseconds

Size in arcseconds — .0881

Gravitational acceleration — 224.93 feet/second^2

Escape velocity — 299.86 miles/second

age — 48 million years old

lifespan—117 million 770 thousand years

Main sequence (prime of its life) lifespan— 106 million years

Time left as a main sequence star —

47 million 990 thousand years

When will this star die?— 69 million 770 thousand years

Fate of the star— planetary nebula, white dwarf star

Peacock is a binary system and is likely a member of the tucana-horologium association of stars that are 45 million years old.

Star's name—
sheliak
beta lyrae

Stellar class— B7(5)/B

Color — blue constellation— lyra

Right ascension— 18 hours 50 minutes 04.79525 seconds

declination— 33 degrees 21 minutes 45.61 seconds

Apparent magnitude— 3.52 (3.25-4.36)

Absolute magnitude— -4.7

distance — 960 light years

mass — 13.16/2.97 suns

radius — 6.0/15.2 suns

luminosity — 6,489/6,500 suns

Surface temperature— 26,300/13,300 degrees kelvin

Velocity of motion — 11.9 miles/second approach

Movement in arcseconds/year—.0019 seconds/year

Direction on the unit circle— 148.57 degrees

parallax—3.39 milliarcseconds

Size—

.02034/.05153 arcseconds

Gravity—324.72 feet/second^2/11 feet 5.1 inches/second^2

Escape velocity—401.47/119.83 miles/ second

age— less than 14 million 325 thousand/23 million years

lifespan— 15 million 917

thousand/657 million 227 thousand years

Main sequence (prime of its life) lifespan— 14 million 325 thousand/592

million 40 thousand years

Time left on main sequence— less than 1 million 592 thousand/ 569 million 41 thousand years

When will this star die?— greater than 1 million 592 thousand/634 million 823 thousand years

fate—supernova, neutron star

*Star's name—
algol
beta persei*

*Stellar class—
B8(5)*

color— blue

constellation— perseus

Apparent magnitude— 2.12

Absolute magnitude— -.07

distance— 90 light years

mass— 3.17 suns

radius— 2.73 suns

luminosity— 182 suns

Surface temperature— 13,000 degrees kelvin

Velocity of motion — 3.7 miles/second

Movement in arcseconds/year — .00166 seconds/year

Direction on the unit circle— 327.74 degrees

parallax—37.27 milliarcseconds

Size in arcseconds — .099

Gravitational acceleration —

378.05 feet/second^2

Escape velocity— 304.33 miles/second

age— 570 million years old (should be less than 503 million years old.)

lifespan— 559 million years Main sequence (prime of its life) lifespan — 503 million years

Time left as a main sequence star — 67 million years

When will this star die?— 11 million years

Fate of the star— planetary nebula, white dwarf star

*Star's name—
vega
alpha lyrae*

*Stellar class—
A0(5)*

color— blue white

constellation— lyra

Apparent magnitude— .026

Absolute magnitude— .582

distance— 25.04 light years

mass— 2.135 suns radius— 2.362 suns luminosity— 40.12 suns

Surface temperature— 9,602 degrees kelvin

Velocity of motion — 8.618 miles/second recession

Movement in arcseconds/year — .353 seconds/year

Direction on the unit circle — 54.13 degrees

parallax — 130.23 milliarcseconds

Size in arcseconds — .3076

Gravitational acceleration — 6.44% of a mile/second^2

Escape velocity—257.73 miles/second

age— 442-468 million years old

lifespan—1 billion 501 million years

Main sequence (prime of its life)

lifespan—1 billion 351 million years

Time left as a main sequence star — 880-910 million years

When will this star die?— 1 billion 30

million-1 billion 60 million years

Fate of the star— planetary nebula, white dwarf

Star's name— sirius

alpha canis majoris

Stellar class— A0(5)

color— blue white

constellation— canis major

Apparent magnitude— -1.47

Absolute magnitude— 1.42

distance— 8.811 light years

mass— 2.063 suns

radius— 1.711 suns luminosity— 25.4 suns

Surface temperature— 9,940 degrees kelvin

Velocity of motion — 3.41 miles/second approach

Movement in arcseconds/year — 1.394 seconds/year

Direction on the unit circle— 182.64 degrees

parallax—379.21 milliarcseconds

Size in arcseconds — .004512

Gravitational acceleration —

11.86% of a mile/second^2

Escape velocity—297.66 miles/second

age— 237-247 million years old

lifespan—1 billion 636 million years

Main sequence (prime of its life) lifespan— 1 billion 472 million years

Time left as a main sequence star — 1 billion 225 million-1 billion 235 million years

When will this star die?— 1 billion 389 million-1 billion 399 million years

Fate of the star— planetary nebula, white dwarf

Sirius will continue to be the brightest star for the next 210,000 years. Over the next 60,000 years, it will increase in brightness, then decrease. The

Sirius binary system is 200-300 million years old. The white dwarf, Sirius b, is 220-238 million years old. 120 million years ago, Sirius b became a white dwarf. It lived 100-118 million years. It started out as a spectral class b main sequence star with mass between 4.99-6.31 suns.

Star's name— castor

alpha geminorum

Stellar class— A1(5)

color— blue white

constellation— gemini

Apparent magnitude— 1.93

Absolute magnitude— .986

distance— 51 light years

mass— 2.76 suns

radius— 2.4 suns

luminosity— 34.49 suns

Surface temperature— 10,286 degrees kelvin

Velocity of motion — 3.722 miles/second recession

Movement in arcseconds/year— .24 seconds/year

Direction on the unit circle— 232.82 degrees

parallax—64.12 milliarcseconds

Size in arcseconds — .1539

Gravitational acceleration — 8.07% of a mile/second^2

Escape velocity — 290.7 miles/second

age— 370 million years old (with mass 2.76 suns, the age should be less than 711 million years.)

lifespan— 790 million years

Main sequence (prime of its life) lifespan— 711 million years

Time left as a main sequence star — 341 million years (could be less than 711 million years.)

When will this star die?— 420 million years (over 79 million years)

Fate of the star— planetary nebula, white dwarf star

Star's name— denebola beta leonis

Stellar class— A3(5)

color— blue white

constellation— leo

Apparent magnitude— 2.113

Absolute magnitude— 1.93

distance— 35.9 light years

mass— 1.78 suns

radius— 1.728 suns

luminosity— 15 suns

Surface temperature— 8,500 degrees kelvin

Velocity of motion — .124 miles/second approach

Movement in arcseconds/year — .128 seconds/year

Direction on the unit circle — 255.68 degrees

parallax—90.91 milliarcseconds

Size in arcseconds — .062

Gravitational acceleration — 10.32% of a mile/second^2

Escape velocity—275.13 miles/second

age— 100-380 million years old

lifespan—2 billion 366 million years

Main sequence (prime of its life)

lifespan— 2 billion 121 million years

Time left as a main sequence star — 1 billion 741 million-2 billion 21 million years

When will this star die?— 1 billion 986

million-2 billion 266 million years

Fate of the star— planetary nebula, white dwarf

Star's name— zosma delta leonis

Stellar class— A4(5)

color— blue white

constellation— leo

Apparent magnitude— 2.56

Absolute magnitude— 1.29

distance— 58.4 light years

mass— 2.2 suns
radius— 2.14 suns
luminosity— 15.5 suns

Surface temperature— 8,296 degrees kelvin

Velocity of motion — 12.524 miles/second approach

Movement in arcseconds/year — .13121 seconds/year

Direction on the unit circle— 313.15 degrees

parallax—55.82 milliarcseconds

Size in arcseconds —

.1195 Gravitational acceleration—

8.09% of a mile/second^2

Escape velocity— 274.86 miles/second

age— 600-750 million years old

lifespan— 1 billion 139 million years

Main sequence (prime of its life) lifespan— 1 billion 254 million years

Time left as a main sequence star — 504 million-654 million years

When will this star die? — 389 million-539 million years

Fate of the star — planetary nebula, white dwarf

Rotation velocity 173.6 miles/

second (oblate spheroid shape)

May be a member of the ursa major moving group, which is about 500 million years old.

Star's name—
altair

alpha aquilae

Stellar class— A7(5)

color— blue white

constellation— aquila

Apparent magnitude— .76

Absolute magnitude— 2.22
distance— 16.73 light years
mass— 1.79 suns
radius— 1.63 suns
luminosity— 10.6 suns

Surface temperature— 6,900-8,500 degrees kelvin

Velocity of motion — 16.13 miles/second approach

Movement in arcseconds/year—

.4372 seconds/year

Direction on the unit circle — 39.66 degrees

parallax — 194.95 milliarcseconds

Size in arcseconds — .3178

Gravitational acceleration — 11.34% of a mile/second^2

Escape velocity— 898.33 miles/second

age— 1 billion 200 million years old

lifespan— 2 billion 333 million years

Main sequence (prime of its life) lifespan— 2 billion 100 million years

Time left as a main sequence star — 900 million years

When will this star die?— 1 billion 133 million years

Fate of the star— planetary nebula, white dwarf star

Shaped like an oblate spheroid with a rotation

velocity of 148.8 miles/ second. Its polar diameter is 20% less than its equatorial diameter.

*Star's name—
porrima
gamma virginis*

Stellar class— F0(5)

color— white

constellation— virgo

Apparent magnitude— 2.74

Absolute magnitude— 2.41

distance— 38.1 light years

mass— 1.56 suns

radius— 2.31 suns

luminosity— 9.29 suns

Surface temperature— 6.757 degrees kelvin

Velocity of motion — 12.09 miles/second approach

Movement in arcseconds/year— .072 seconds/year

Direction on the unit circle— 173.67 degrees

parallax—85.58 milliarcseconds

Size in arcseconds — 1.977

Gravitational acceleration — 4.92% of a mile/second^2

Escape velocity — 222.77 miles/second

age— 1 billion 140 million years old

lifespan— 3 billion 290 million years

Main sequence (prime of its life) lifespan— 2 billion 961 million years

Time left as a main sequence star — 1 billion 821 million years

When will this star die? — 2 billion 150 million years

*Fate of the star—
planetary nebula,
white dwarf star*

Star's name—
talitha
iota ursae majoris

*Stellar class—
F0(5)*

color— white

*constellation—
ursa major*

*Apparent
magnitude— 3.14*

Absolute magnitude— 2.31
distance— 47.3 light years
mass— 1.7 suns
radius— 3.02 suns
luminosity— 9.87 suns

Surface temperature— 6,000 degrees kelvin

Velocity of motion — 5.5 miles/ second recession

Movement in arcseconds/year— .235 seconds/year

Direction on the unit circle— 241.11 degrees

parallax—68.92 milliarcseconds

Size in arcseconds — .208

Gravitational acceleration — 3.14% of a mile/second^2

Escape velocity — 203.39 miles/second

age— 620 million years old
lifespan— 2 billion 654 million years

Main sequence (prime of its life) lifespan— 2 billion 389 million years

Time left as a main sequence star — 1 billion 769 million years

When will this star die? — 2 billion 34 million years

Fate of the star—planetary nebula, white dwarf star

Star's name—
procyon

alpha canis minoris

Stellar class—
F5(5)
color— white

*constellation—
ursa minor*

Apparent magnitude— .34

Absolute magnitude— 2.66

distance— 11.46 light years

*mass— 1.499 suns
radius— 2.048 suns luminosity— 6.93 suns*

Surface temperature— 6,530 degrees kelvin

Velocity of motion — 1.984 miles/second approach

Movement in arcseconds/year — 1.274 seconds/year

Direction on the unit circle — 298.42 degrees

parallax — 284.56 milliarcseconds

Size in arcseconds — .5828

Gravitational acceleration — 5.15% of a mile/second^2

Escape velocity— 231.92 miles/second

age — 1 billion 740 million-2 billion years old

lifespan — 3 billion 635 million years

Main sequence (prime of its life) lifespan — 3 billion 271 million years

Time left as a main sequence star — 1 billion 274 million-1 billion 531 million years

When will this star die?— 1 billion 635 million-1 billion 895 years

*Fate of the star—
planetary nebula,
white dwarf star*

Star's name— sol (our sun)

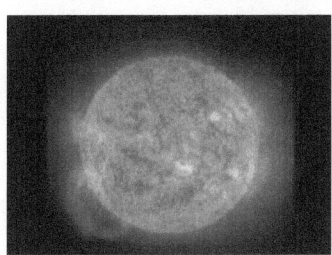

Stellar class— G2(5)

color— yellow

constellation— n/a

Apparent magnitude— -26.74

Absolute magnitude— 4.83

distance—93 million miles from earth

mass— 1.989×10^{30} kilograms

radius— $693{,}979.2344$ kilometers

luminosity— 3.828×10^{26} watts

Surface temperature— 5,880 degrees kelvin

Velocity of motion — 12.4 miles/second towards lambda hercules

Movement in arcseconds/year— .5298 seconds/year

Direction on the unit circle— 202.78 degrees

parallax— 2,129,040 milliarcseconds

Size in arcseconds — 2,129.04

Gravitational acceleration — 17.03% of a miles/second^2

Escape velocity—382.974 miles/second

age— 4 billion 568 million 200 thousand years old

lifespan—10 billion years

Main sequence (prime of its life) lifespan— 9 billion years

Time left as a main sequence star —4 billion 397 million years.

In 1 billion years, the sun will be 10% brighter and the earth's surface temperature will be 116 degrees fahrenheit. In 3.5-4.5 billion years, the sun will be 35-40% brighter than it is now. The sun will become a red giant star in 5 billion years.

When will this star die?— 8 billion years (will become a white dwarf)

Fate of the star—planetary nebula, white dwarf star (mass .5405 suns)

NOTE—during its red giant lifetime, the sun will expand to 256 times larger than it is now. It will have a luminosity of 34,265 suns and an absolute magnitude of -6.51, apparent magnitude of -38.01 if it were where the sun is with respect to us, and a surface temperature of 5,000 kelvin. Sol began its life 40 million years before earth formed. It has increased its luminosity 1% every 100 million years (30% in 4.5 billion years). In 1.1 billion years, the sun will be 10% brighter, and in 3.5 billion years, it will be 40% brighter. the helium fusing phase will last for 130 million years, and the sun will expand past the orbits of mercury and venus. in 7.59 billion years, the sun will become a red giant. Its mass will then be decreased 67% and will be 1/2 as hot as it is now. the earth will be engulfed before the red giant phase and the sun will grow .25 AU during the next 500,000 years.

*Star's name—
rigel kent
alpha centauri*

*Stellar class—
G2(5)*

color— yellow

constellation— centaurus

Apparent magnitude— .01

Absolute magnitude— 4.38

distance— 4.367 light years

*mass— 1.1 suns
radius— 1.2234
suns luminosity—
1.519 suns*

Surface temperature— 5,790 degrees kelvin

Velocity of motion — 13.268 miles/second approach

Movement in arcseconds/year — .2374 seconds/year

Direction on the unit circle— 97.5 degrees

parallax—20.23 milliarcseconds

Size in arcseconds — .9334

Gravitational acceleration — 12.37% of a mile/second^2

Escape velocity — 257.05 miles/second

age — 4 billion 820 million years old

lifespan— 7 billion 880 million years

Main sequence (prime of its life) lifespan— 7 billion 92 million years

Time left as a main sequence star — 2

billion 692 million years

When will this star die?— 3 billion 480 million years

Fate of the star— planetary nebula, white dwarf star

Star's name— alpha centauri B

Alpha Centauri

α Centauri AB is the bright star to the left, with Proxima Centauri circled in red. The bright star to the right is β Centauri

Stellar class— K1(5)

color— orange constellation— centaurus

Apparent magnitude— 1.33 (1.35)

Absolute magnitude— 5.71 (5.57)

distance— 4.367 light years (4.4 light years)

mass— .907 suns
radius— .8632 suns (.88 suns)

luminosity— .445 suns (.58)

Surface temperature— 5,260 (5,052) degrees kelvin

Velocity of motion — 13.268 miles/second approach

Movement in arcseconds/year—.2374 seconds/year

Direction on the unit circle— 97.5 degrees

parallax—20.23 milliarcseconds

Size in arcseconds — .01746

Gravitational acceleration — 20.49% of a mile/second^2

Escape velocity— 277.87 miles/second

age— 6 billion 500 million years old

lifespan— 12 billion 764 million years

Main sequence (prime of its life) lifespan— 11

billion 488 million years

Time left as a main sequence star — 4 billion 988 million years

When will this star die?— 6 billion 264 million years

*Fate of the star—
planetary nebula,
white dwarf star*

SUBGIANT STARS

Star's name—

acrux

alpha crucis

Stellar class— B.5(4)

color— blue

constellation— crux

Apparent magnitude— .76

Absolute magnitude— -2.2

distance— 320 light years

mass— 17.8 suns

radius— greater than 6.6 suns

luminosity— 648.84 suns (should 25,000)

Surface temperature— 24,000 degrees kelvin

Velocity of motion — 6.944 miles/ second approach

Movement in arcseconds/year— .03879 seconds/ year

Direction on the unit circle— 244.97 degrees

parallax—10.13 milliarcseconds

Size in arcseconds — .066858

Gravitational acceleration — 6.88% of a mile/second^2

Escape velocity— 718.03 miles/second

age— less than 6 million 700 thousand years old

lifespan—less than 7 million 480 years

Main sequence (prime of its life) lifespan— less

than 6 million 700 thousand years

Time left as a main sequence star — star has already completed this phase of its lifetime

When will this star die? — less than

748 thousand years

Fate of the star— supernova, neutron star

Star's name— alfirk beta cephei

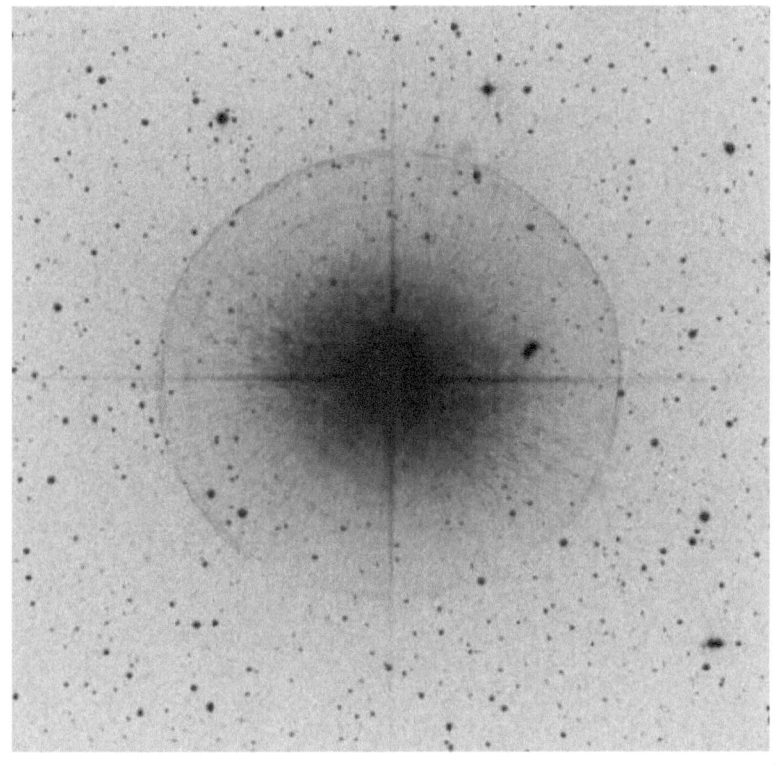

Stellar class— B1(4)

color— blue

constellation— cepheus

Apparent magnitude— 3.16 to 3.27

Absolute magnitude— -3.03
distance— 690 light years
mass— 12.2-19.5 suns
radius— 5.6 suns
luminosity— 15,100 suns

Surface temperature— 27,000 degrees kelvin

Velocity of motion — 5.08 miles/second approach

Movement in arcseconds/year— .00839 seconds/year

Direction on the unit circle— 37.54 degrees

parallax—4.76 milliarcseconds

Size in arcseconds — .0267

Gravitational acceleration — 6.55-10.47% of a mile/ second^2

Escape velocity— 525.85 miles/ second

age— 8.7 million years old

lifespan—less than 5 million 955 thousand-19 million 235 thousand years

Main sequence (prime of its life)

lifespan— less than 5 million 360 thousand-17 million 312 thousand years

Time left as a main sequence star — star has already

completed this phase of its lifetime

When will this star die?— less than 596 thousand years

Fate of the star— supernova, neutron star

Star's name— algenib gamma pegasi

Stellar class— B2(4)

color— blue

constellation— pegasus

Apparent magnitude— 2.84

Absolute magnitude— -2.64

distance— 390 light years

mass— 8.9 suns
radius— 4.8 suns
luminosity— 5,840 suns

Surface temperature— 21,179 degrees kelvin

Velocity of motion — 2.542 miles/second recession

Movement in arcseconds/year — .00928 seconds/year

Direction on the unit circle— 273.38 degrees

parallax—8.33 milliarcseconds

Size in arcseconds — .03998

Gravitational acceleration — 6.514% of a mile/second^2

Escape velocity — 358.61 miles/second

age — 15.5-21.9 million years old

lifespan—less than 48 million 900 thousand years

Main sequence (prime of its life) lifespan— less than 44 million years

Time left as a main sequence star — star has already completed this phase of its lifetime

When will this star die?— less than 27 million-33

*million 400
thousand years
Fate of the star—
supernova, white
dwarf*

*Star's name—
lesath*

upsilon scorpi

Stellar class— B2(4)

color— blue

constellation— scorpius

Apparent magnitude— 2.7

Absolute magnitude— -3.53

distance— 577.3 light years

mass— 11.4 suns

radius— 6.1 suns

luminosity— 7,381 suns (should be 12,300)

Surface temperature— 22,831 degrees kelvin

Velocity of motion — 4.96 miles/second recession

Movement in arcseconds/year — .03 seconds/year

Direction on the unit circle — 85 degrees

parallax—5.66 milliarcseconds

Size in arcseconds — .0345

Gravitational acceleration — 5.16% of a mile/second^2

Escape velocity—370.58 miles/second age—17.4-22.6 million years old

lifespan—less than 22 million 800 thousand years

Main sequence (prime of its life) lifespan— less than 20 million 500 thousand years

Time left as a main sequence star — star has already

completed this phase of its lifetime

When will this star die?— less than 2 million 50 thousand years (could dies any time)

Fate of the star— supernova, neutron star

Star's name— taygeta

19 Tauri

19 Tauri in the Pleiades cluster

Stellar class— B6(4)

color— blue

constellation— taurus

Apparent magnitude— 4.3

Absolute magnitude— -1.35

distance— 440.3 light years

mass— 4.5 suns

radius— 3.71 suns

luminosity— 600 suns

Surface temperature— 9,000 degrees kelvin

Velocity of motion — 6.262 miles/second

Movement in arcseconds/year— .0406 seconds/year

Direction on the unit circle— 320.71 degrees

parallax—7.97 milliarcseconds

Size in arcseconds — .02957

Gravitational acceleration —

5.5% of a mile/second^2

Escape velocity— 481.53 miles/second

age— less than or equal to 1 billion years old

lifespan—less than 209 million 500 thousand years

Main sequence (prime of its life) lifespan— less than 232 million 800 thousand years

Time left as a main sequence star — star has already completed this phase of its lifetime

When will this star die?— less than 20 million 950 thousand years

Fate of the star— planetary nebula, white dwarf star

*Star's name—
phact
alpha colombae*

Stellar class— B7(4)

color— blue

constellation— colomba

Apparent magnitude— 2.645

Absolute magnitude— -1.87
distance— 261 light years
mass— 4.5 suns
radius— 5.8 suns
luminosity— 478.78 suns

Surface temperature— 12,963 degrees kelvin

Velocity of motion — 21.7 miles/second recession

Movement in arcseconds/year— .0248 seconds/year

Direction on the unit circle— 174.05 degrees

parallax—12.48 milliarcseconds

Size in arcseconds — .0724

Gravitational acceleration — 2.25% of a mile/second^2

Escape velocity— 238.78 miles/second

age— less than 209 million 500 thousand years old

lifespan—less than 232 million 790 thousand years

Main sequence (prime of its life) lifespan— less

than 209 million 500 thousand years

Time left as a main sequence star — star has already completed this phase of its lifetime

When will this star die?— less than 23 million 300 thousand years

Fate of the star— planetary nebula, white dwarf

*Star's name—
regulus
alpha leonis*

Stellar class— B8(4)

color— blue

constellation— leo

Apparent magnitude— 1.4

Absolute magnitude— -.57

distance— 79.3 *light years*

mass— 3.8 *suns*
radius— 3.092 *suns*
luminosity— 288 *suns*

Surface temperature— 12,460 degrees kelvin

Velocity of motion — 3.658 miles/ second recession

Movement in arcseconds/year—

.00608 seconds/year

Direction on the unit circle— 178.57 degrees

parallax—41.11 milliarcseconds

Size in arcseconds — .127

Gravitational acceleration — 6.69% of a mile/second^2

Escape velocity — 300.52 miles/second

age— less than or equal to 1 billion years old

lifespan—less than 1 billion 100 million years

Main sequence (prime of its life) lifespan— less than 1 billion years

Time left as a main sequence star — star has already completed this phase of its lifetime

When will this star die?— less than 100 million years

Fate of the star— planetary nebula, white dwarf star

Star's name— alpheratz

alpha andromedae

Stellar class— B8(4)

color— blue

constellation— Andromeda

Apparent magnitude— 2.06

Absolute magnitude— -.19
distance— 97 light years
mass— 3.8 suns
radius— 2.7 suns

*luminosity—
101.88 suns*

*Surface temperature—
13,800 degrees kelvin*

*Velocity of motion
— 6.572 miles/
second approach*

Movement in arcseconds/year—.1644 seconds/year

Direction on the unit circle— 355.26 degrees

parallax—33.62 milliarcseconds

Size in arcseconds — .0908

Gravitational acceleration — 8.77% of a mile/second^2

Escape velocity — 321.6 miles/second

age— 60 million years old lifespan —greater than 355 million years

Main sequence (prime of its life) lifespan— greater than 319 million

700 thousand years

Time left as a main sequence star — according to its mass, this star should still be a main sequence star.

When will this star die?— less than 319 million 700 thousand years

Fate of the star— supernova, neutron star

*Star's name—
algorab
delta corvi*

*Stellar class—
A0(4)*

color— blue white

constellation— corvus

Apparent magnitude— 2.962

Absolute magnitude— .2

distance— 86.76 light years

mass— 2.74 suns
radius— 2.66 suns
luminosity— 69 suns

Surface temperature— 10,400 degrees kelvin

Velocity of motion — 5.58 miles/second recession

Movement in arcseconds/year — .1418 seconds/year

Direction on the unit circle — 142.95 degrees

parallax — 37.55 milliarcseconds

Size in arcseconds — .09988

Gravitational acceleration — 6.52% of a mile/second^2

Escape velocity— 286.63 miles/second

age— 236-274 million years old

lifespan—less than 804 million 700 thousand years

Main sequence (prime of its life) lifespan— less than 724 million 200 thousand years

Time left as a main sequence star — according to its mass, this star should sell be a main sequence star.

When will this star die?— less than

530 million 700 thousand-568 million 700 thousand years

Fate of the star— planetary nebula, white dwarf star

*Star's name—
sabik*

eta ophiucus

*Stellar class—
A1(4)*

color— blue white

constellation— ophiucus

Apparent magnitude— 2.43

Absolute magnitude— .27

distance— 88 light years

mass— 2.966 suns

radius— 7.45 suns

luminosity— 66.47 suns

Surface temperature— 6,150 degrees kelvin

Velocity of motion — 2.966 miles/second

Movement in arcseconds/year — .09925 seconds/year

Direction on the unit circle— 75.52 degrees

parallax—36.91 milliarcseconds

Size in arcseconds — .27498

Gravitational acceleration — 47

feet 5.76 inches/second^2

Escape velocity— 275.88 miles/second

age— less than 594 million years old

lifespan—less than 660 million years

Main sequence (prime of its life) lifespan— less than 594 years

Time left as a main sequence star — star has already

completed this phase of its lifetime

When will this star die? — less than 66 million years

Fate of the star — planetary nebula, white dwarf star

Sabik is the north pole star of the planet uranus.

*Star's name—
menkalanin
beta aurigae*

Stellar class— A1(4)

color— blue white

constellation— auriga

Apparent magnitude— 1.9

Absolute magnitude— .55

distance— 81.1 light years

mass— 2.389 suns

radius— 2.77 suns

luminosity— 48 suns

Surface temperature— 9,350 degrees kelvin

Velocity of motion — 11.284 miles/second approach

Movement in arcseconds/year— .00095 seconds/year

Direction on the unit circle— 268.39 degrees

parallax—40.21 milliarcseconds

Size in arcseconds — .11138

Gravitational acceleration — 5.24% of a mile/second^2

Escape velocity— 251.75 miles/second

age— 370 million years old

lifespan— less than 1 billion 113 million years

Main sequence (prime of its life) lifespan— less

than 1 billion 20 million years

Time left as a main sequence star — has left the main sequence

When will this star die? — less than

563 million 500 thousand years

Fate of the star—planetary nebula, white dwarf star

*Star's name—
Alhena
gamma
geminorum*

Stellar class— A1.5(4)

color— blue white constellation— gemini Apparent magnitude— 1.915

Absolute magnitude— -.68

distance— 110.9 light years

mass— 2.81 suns

radius— 3.3 suns

luminosity— 123 suns

Surface temperature— 9,260 degrees kelvin

Velocity of motion — 7.75 miles/second approach

Movement in arcseconds/year—.05497 seconds/year

Direction on the unit circle— 357.9 degrees

parallax—29.84 milliarcseconds

Size in arcseconds — .0985

Gravitational acceleration — 4.34% of a miles/second^2

Escape velocity — 250.15 miles/second

age— less than 680 million years old

lifespan—less than 756 million years

Main sequence (prime of its life) lifespan— less

than 680 million years

Time left as a main sequence star — star has already completed this phase of its lifetime

When will this star die? — less than

75 million 550 thousand years

Fate of the star—planetary nebula, white dwarf star

This star is a spectroscopic binary

system with a period of orbit of 4,614.51 days in a highly keplerian orbit with a separation distance of 5.42 AU (504.39 million miles).

*Star's name—
acamar
theta eridani*

*Stellar class—
A3(4)*

color— blue white

constellation— eridanus

Apparent magnitude— 3.3

Absolute magnitude— -.59

distance— 161 light years

mass— 2.6 suns
radius— 16 suns
luminosity— 145 suns

Surface temperature— 8,200 degrees kelvin

Velocity of motion — 7.378 miles/second

Movement in arcseconds/year — .022 seconds/year

Direction on the unit circle— 154.93 degrees

parallax—20.23 milliarcseconds

Size in arcseconds — .3237

Gravitational acceleration — 9 feet/ second^2

Escape velocity— 209.28 miles/ second

age— less than 491 million years old

lifespan — less than 546 million years

Main sequence (prime of its life) lifespan — less than 491 million years

Time left as a main sequence star — star

has already completed this phase of its lifetime

When will this star die?— less than 55 million years

Fate of the star— planetary nebula, white dwarf star

This star is a binary system with a separation distance of 771.9 million miles and an orbital period of 23.91 years. Its companion is an A1 stars, apparent magnitude of 4.3, absolute magnitude of . 83, luminosity of 36 suns, temperature *of 9,200 kelvin, and mass of 2.4 suns.*

*Star's name—
ruchbah
delta cassiopeiae*

*Stellar class—
A5(4)*

color— blue white

constellation— cassiopeia

Apparent magnitude— 2.68

Absolute magnitude— .28

distance— 99.48 light years

mass— 2.49 suns

radius— 3.9 suns

luminosity— 72.88 suns

Surface temperature — 7.980 degrees kelvin

Velocity of motion — 4.154 miles/second approach

Movement in arcseconds/year— .0513 seconds/year

Direction on the unit circle— 349.53 degrees

parallax—32.81 milliarcseconds

Size in arcseconds — .12796

Gravitational acceleration — 2.755% of a mile/second^2

Escape velocity — 225.66 miles/second

age— 600 million years old

lifespan—less than 1 billion 22 million years

Main sequence (prime of its life) lifespan— less

than 920 million years

Time left as a main sequence star — this star should still be a main sequence star according to its mass.

When will this star die?— less than 422 million years

Fate of the star— planetary nebula, white dwarf star

This star system is an eclipsing binary

system. The 2 stars are separated by 151.37 million miles and have a period of 759 days. The main star is 4% beyond its main sequence lifetime. An excess of infrared emissions at wavelength 60 micrometers suggests the presence of a debris disk. The heat of the debris *disk is 85 degrees kelvin which corresponds to an orbital radius of 88 AU (8.184 billion miles). Our star's copier belt has an orbit of 30-50*

AU (2.79-4.65 billion miles).

www.ingramcontent.com/pod-product-compliance
Lightning Source LLC
Chambersburg PA
CBHW030605220526
45463CB00004B/1178